EXPLORING HI-TECH JOBS

Hi-Tech Jobs in
ROBOTICS

Stuart A. Kallen

San Diego, CA

© 2024 ReferencePoint Press, Inc.
Printed in the United States

For more information, contact:
ReferencePoint Press, Inc.
PO Box 27779
San Diego, CA 92198
www.ReferencePointPress.com

LIBRARY OF CONGRESS CATALOGING-IN-PUBLICATION DATA

Names: Kallen, Stuart A., 1955- author.
Title: Hi-tech jobs in robotics / by Stuart A. Kallen.
Description: San Diego, CA : ReferencePoint Press, Inc., 2024. | Series:
  Exploring hi-tech jobs | Includes bibliographical references and index.
Identifiers: LCCN 2023032229 (print) | LCCN 2023032230 (ebook) | ISBN
  9781678207069 (library binding) | ISBN 9781678207076 (ebook)
Subjects: LCSH: Robotics--Vocational guidance--Juvenile literature.
Classification: LCC TJ211.25 .K352 2024 (print) | LCC TJ211.25 (ebook) |
  DDC 629.8/92023--dc23/eng/20230831
LC record available at https://lccn.loc.gov/2023032229
LC ebook record available at https://lccn.loc.gov/2023032230

Contents

Introduction: Rapid Growth in Robotics

In 2023 the Los Angeles Police Department put a new type of canine into service. It is a doglike robot called Spot. While Spot is commonly called a robodog, it is officially referred to as a Legged Squad Support System (LS3) and was manufactured by Boston Dynamics. Spot is 28 inches (71 cm) high and weighs 70 pounds (31.8 kg)—about the same size as a dalmatian, the white-and-black dog breed conventionally associated with fire stations.

Spot can map terrain with several 360-degree, heat-sensing, night vision cameras. The robodog processes data, creates 3-D imagery, and transmits real-time information to police officers who control the robodog with a tablet computer. Spot also responds to ten voice commands. The $280,000 robodog climbs stairs, opens doors, and employs sensors to detect chemicals, radiation, and explosives. In addition to police work, LS3s are employed for a variety of tasks by utility and construction companies, manufacturing businesses, and the military.

Like all robots, Spot responds to programmed commands and can perform complex tasks with little or no human interaction. The machine is one of a rapidly expanding legion of robots that combine science, technology, and engineering into three basic robotic components. The digital "brains" of robots, called computational units, are made up of computer chips and software, usually running artificial intelligence (AI) programs. The units "see and hear" using a variety of sensors and cameras that gather information from the real-world environment. Robots achieve physical movement with mechanical "muscles" called actuators. This hardware consists of motors, pistons, grippers, wheels, and gears that propel robots and allow them to grab, lift, and perform other tasks. The actuators are driven by power sources that might include a combination of batteries, pneumatic (air) systems, and hydraulic (fluid) systems.

Robodogs tend to attract media attention, but robotics is so well integrated into modern society as to be nearly invisible. Robots assemble vehicles, electronics, and a wide range of consumer goods. Robotic machinery is used in self-driving cars, trucks, airplanes, and agricultural equipment. Robots play a central role in the production of pharmaceuticals, while surgical robots help save lives every day. And in the not-too-distant future, personal robots (telepresence robots) will perform household tasks and care for the sick and the elderly.

Jobs in Robotics

The career field for robotics is experiencing unprecedented growth due to advances in engineering, AI, and machine learning. This hi-tech progress is converging to produce the next generation of automated machines that are increasingly smaller, cheaper, safer, and more easily programmed.

A 2020 report by the World Economic Forum says the revolution in robotics is expected to create 97 million new jobs by the end of the decade. Scientists and engineers skilled in mechanics, robotics, AI, computing, and electronics are needed to design, test, and build robots. Technicians, programmers, and others who specialize in robotic processes are in demand anywhere robots are at work.

The e-commerce company Amazon is at the forefront of robotic innovation. The company is building a new class of robots with advanced computer vision. The robots move freely through cluttered warehouse fulfillment centers that are as large as twenty-five football fields. These autonomous, or self-directed, robots are controlled by cloud-based algorithms, or rules that govern specific processes, that allow them to transport large and unwieldy items around obstacles that might include pallets, shelves, and people. The robots rely on what Amazon scientists call semantic understanding. This is the ability for a robot to see the world in three dimensions and recognize each object that it encounters. According to Siddhartha Srinivasa, director of Amazon Robotics AI, "The way a robot moves around a trash can is probably going to be different from the way it navigates around a person or a precious asset. The only way the robot can know that is if it's able to identify, 'Oh that's the trash can or that's the person.' And that's what our AI is able to do."[1]

A Promising Future

As major corporations like Amazon move to increase their use of robots to fulfill orders, those who specialize in robotics are in high demand. According to the International Federation of Robotics, new robot installations worldwide skyrocketed to a record level of more than 517,000 units in 2022, a growth rate of 31 percent over 2020. With this spectacular increase, robotics specialists are earning from $85,000 to as much as $136,000 annually, more on average than any other tech workers. And according to the Bureau of Labor Statistics, robotics and AI job postings are surging, with double- and triple-digit increases seen every year. Because the global robotics market is expected to attract nearly $150 billion in revenue by 2025, a career in robotics is a career with a future.

Robotics Programmer

What Does a Robotics Programmer Do?

Robots are machines made up of an intricate mix of gears, pistons, switches, actuators, sensors, and casing. But these metal, plastic, and rubber components would not function without lines of code written by computer programmers who specialize in robotics. Product manager Sam Daley writes, "Without a set of code telling it what to do, a robot would just be another piece of simple machinery. Inserting a program into a robot gives it the ability to know when and how to carry out a task."[2]

Robot programmers, sometimes called roboticists, design, write, test, debug, and maintain the source code of computer programs used by robots. Their work allows sensor-rich robots to learn ever-more-complicated tasks. The job requires robotics programmers to understand artificial intelligence (AI), data abstraction, user interface design, and digital security. These programming concepts and topics are important to robotics, but they are also basic to most types of computer programming.

Code created by robotics programmers enables a robot to perceive its environment, make decisions, and

A Few Facts

Number of Jobs
About 1.62 million in 2021*

Pay
Median annual salary of
$105,047*

Educational Requirements
Bachelor's degree in computer science, software engineering, or a related field

Personal Qualities
Good at math and computer programming, concentration, analytical, strong communication skills

Work Settings
Indoors in offices and on factory floors

Future Job Outlook
Growth rate of 25 percent through 2031*

* For all software developers, including robotics programmers

carry out tasks. Deep learning algorithms created by programmers help a robot detect objects, avoid collisions, and choose a route using sensors, actuators, and cameras. Robotics software programs also connect and coordinate core computing components such as central processing units, graphics processing units, and microcontrollers.

Programming for communication is necessary, since many types of robots need the ability to link with one another and interpret verbal commands given by humans. But beyond communicating, robots need to learn how to do their jobs better and even adjust to changes in their environments. Thus, a roboticist must understand the branch of AI called machine learning. According to programmer Shanmukh Rock, "Programming is crucial in enabling robots to learn from their experiences and improve their performance over time. This could involve programming machine learning algorithms that enable the robot to adapt to new situations and improve its performance based on past experiences."[3]

Writing code and applications for robotic equipment in a simulated computer environment is called offline programming; when the virtual robot within the program can carry out its tasks correctly, the commands are uploaded to the real robot. A second method, called online programming, involves physically moving a robot through a series of actions using what is called a teach pendant. The teach pendant is a control that looks something like a handheld gaming device with buttons, dials, and a touch screen. To program a robot's actions, the operator moves it from place to place using buttons on the pendant. Each motion or position of the robot can be saved on the pendant. In this case the actions are translated into digital commands. After the entire range of movements has been stored, the robot can autonomously re-create the actions at a rapid rate by following that "learned" routine. This is the most common form of robotics programming.

A Typical Workday

Robotics programmers begin their days going through emails from team members who need guidance or who have discovered problems with an application. When writing code, programmers need to remain focused on minute details for long periods. On a typical workday a robotics programmer might test, maintain, and monitor existing computer programs, develop new software, and review and scrutinize computer printouts to locate problems in performance. When a company wants to bring a new robotic product to market, programmers are involved in every step of the process. They spend their days working on product development, prototype testing, and production supervision and debugging software programs when they are near completion.

Education and Training

Some robotics programmers began their careers as kids, building and programming their own robots. Roy Barazani began building robots from kits when he was in kindergarten and went on to found a robot toy company called the OffBits. Barazani says, "The hands-on building process allows users to learn about robotics and engineering concepts in a fun and interactive way. The step-by-step instructions for building and programming [robots makes] it easy for users of all ages to understand the principles behind the technology."[4]

While building toy robots is fun, those with an interest in robotics programming need to attend to their formal education. Most programmers excel at math, including calculus, trigonometry, and algebra. It is also important to take computer science courses that emphasize programming languages. The two most popular coding languages for robotics are Python and C/C++. Others include Lisp and Pascal. Motivated students can teach themselves the basics of programming. There are numerous free courses available online, along with podcasts and tech blogs, that provide tutorials, workshops, and community resources. Software engineer Chris Loy says, "My main piece of advice is simply to get coding! . . . Look for articles that have code in them or link to a GitHub repository and then play around with the code."[5] Many newbie coders find coding boot camps to be extremely helpful. These twenty-four-week courses, which are usually open to high school students, teach coding and career development.

In addition to understanding coding languages, robotics programmers need to be skilled in the use of two software applications. Robot Operating System (ROS) is a set of software tools and libraries for obtaining, building, writing, and running code across multiple computers. ROS is open source, which means it can be downloaded for free. A second open-source programming platform, MATLAB, is used to analyze data and develop algorithms for machine learning applications.

Most robotics programmers pursue a bachelor of computer science degree in college. They take math-heavy courses in computer programming, algorithms, data structures, logic and computation, calculus, linear algebra, and statistics. Other courses focus on operating systems, real-time computing, AI, machine learning, and software testing.

Many companies offer summer intern positions to computer science students who are studying robotics programming. Can-

Robotics programmers design, write, test, debug, and maintain the source code of computer programs that are essential to robotic function. The work of these programmers enables sensor-rich robots to perform complicated tasks.

didates work with teams in a fast-moving environment to integrate, deploy, and support complex software systems. Interns participate in research and development, problem solving, maintenance, and other tasks. To qualify for internships, candidates are required to be enrolled in a college or university, and they must be familiar with various robotics operating systems, installation procedures, coding, and process management.

Skills and Personality

Robotics programmers need strong math and programming skills so that writing code becomes second nature. The ability to concentrate on extremely complex lines of code for long periods is also necessary, along with an eye for detail; even the slightest error can result in a robot performing poorly or breaking down. An analytical mindset is useful for problem solving.

Good communication skills are a must because robotics programmers work with developers, engineers, managers, and

Be a Problem Solver

"Certain programming languages are better than others for different use cases. But at their core, all modern programming languages are simply a mode of communication between humans and computers. Proficient programmers can write code to fully utilize the language. But the real skill comes from knowing how to solve the problem at hand. This problem solving ability is what sets programmers up for success in the long run. . . . That's why the main piece of advice I like to share with new coders is this: becoming a great problem solver will enable you to become a great programmer."

—Fahim ul Haq, software programmer

Fahim ul Haq, "The 1 Piece of Coding Advice I Wish I Had Received," Educative, February 22, 2023. www.educative.io.

others. Programmers should be able to express complex ideas clearly in plain language and listen closely to the comments, suggestions, and questions posed by others. Robotics blogger Devin Partida adds, "Good communication skills will also help you with networking. In the sciences, networking can help you meet people with knowledge and ideas that might help you solve complex problems."[6]

Good communication extends to writing. Robotics programmers are required to publish weekly progress reports and other documents. Projects can take many weeks or months to finish. Thus, the old saying "patience is a virtue" is especially true for robotics programmers. Robotics programmers need to be calm and supportive while working on what might seem like a never-ending cycle of programming, testing, and refining.

Despite the lengthy process, successful robotics programmers need to adapt rapidly to change. Employers may want to add another skill to a robot's repertoire, and the programmer must be ready to accommodate. Furthermore, technology is moving quickly, and programmers need to regularly update their skills and learn new concepts and techniques.

Working Conditions

Robotics programmers work forty hours a week in offices alongside managers, research scientists, and engineers who specialize in sales, manufacturing, and robotics. Some spend time in labs programming prototypes as they are developed. Those who specialize in manufacturing robots work on busy factory floors, where they might be required to wear safety equipment such as hard hats, goggles, and ear protection.

Employers and Earnings

Robotics programmers work in a variety of industries dominated by automation and robotics. Robotics programmer jobs are being offered by online retailers, aerospace companies, makers of pharmaceuticals and medical equipment, and manufacturers of robotic assembly lines. Next-generation robots are building electric vehicles for all major automobile companies. Construction equipment companies are creating robots to paint, dig, and even lay bricks. In agriculture, John Deere is working to develop autonomous tractors that spray fertilizers and pick crops using machine vision and AI. There is a strong demand for workers to fill these hi-tech jobs, and companies are offering good salaries to those with the necessary skills. The Bureau of Labor Statistics (BLS) defines robotics programmers as software developers. The jobs website ZipRecruiter says the average annual pay for software developers in 2023 was $105,047.

Future Outlook

According to the BLS, demand for all software developers will experience exceptional growth, increasing by 25 percent through 2031. Because robotics skills are in high demand, programmers will likely benefit from that growth.

Find Out More

Codecademy

www.codecademy.com

This online school offers free coding lessons in numerous programming languages, including Python, C++, Git, JavaScript, and CSS. Students can sign up and begin coding within minutes.

National Robotics Education Foundation (NREF)

www.the-nref.org

The NREF is dedicated to the goal of creating robotics education curriculum for elementary, middle, and high schools. The foundation's website provides curricula and lesson programs, a robotics education journal, and links to related robotics websites.

TopCoder

www.topcoder.com

This website, with nearly 1 million highly skilled members, hosts bimonthly computer programming contests in which software developers, designers, and student programmers compete for cash prizes while solving coding problems.

Manufacturing Engineer

What Does a Manufacturing Engineer Do?

The work of manufacturing engineers can be seen in most rooms in a modern home. Almost every item—including food, toys, medicine, appliances, TVs, and smartphones—is produced on automated production lines overseen by manufacturing engineers. As a manufacturing engineer known as Dominque writes on the Society of Women Engineers website, "There's not something that you can wake up and touch today that is not manufactured . . . [including] automobiles, electrical devices, or makeup. It's all manufactured."[7]

Industrial robots play a central role in assembly lines. Robotic arms, welders, paint sprayers, circuit board printers, conveyor belts, and other equipment are all operated by computers that manage the intricate steps of creating products. Manufacturing engineers—sometimes called production engineers, process engineers, or industrial engineers—are needed to design, build, and test these complex production systems. The engineers plan manufacturing operations, set up assembly lines, monitor production quality, and develop machinery safety protocols.

A Few Facts

Number of Jobs
About 301,000 in 2021*

Pay
Median annual salary of $95,300*

Educational Requirements
Bachelor of science degree in engineering

Personal Qualities
Strong operational skills, knowledge of computers and robotics, problem solver, team player

Work Settings
Shift work in offices and factory floors

Future Job Outlook
Growth rate of 10 percent through 2031*

* For all industrial engineers, including manufacturing engineers

Manufacturing engineers create software programs that guide the actions of the entire assembly line. Each machine must do exactly what it is supposed to do in coordination with other machinery, and the machines must work in harmony twenty-four hours a day while repeatedly producing thousands of identical products in a seamless, efficient manner. This is demanding work. As manufacturing engineer Becky Miller explains, each machine must move very precisely, or the production line will break down: "An error the size of a human hair can make a difference."[8]

Engineers who specialize in manufacturing spend time researching and testing new equipment that might make their assembly line more efficient and cost effective. If a new machine or system looks promising, the engineer will conduct what is called a return on investment analysis. He or she evaluates the machinery, determines its quality and efficiency, and decides whether the benefits outweigh the purchase costs.

In addition to their other duties, manufacturing engineers fill out forms and production reports. They create and review schedules, engineering specifications, safety records, environmental requirements, and budgets. When paperwork reveals glitches, such as a high number of rejected products, manufacturing engineers devise and test new quality control procedures to minimize problems. Joe Dyer, a manufacturing engineer who oversees production of assembly-line equipment, describes the diversity of the tasks he attends to: "On any given day in the life of an engineer in manufacturing, I am expected to provide expertise in: manufacturing processes, quality tools, industrial engineering, automation technologies, design, . . . process development, and continuous improvement."[9]

A Typical Workday

The workday of a manufacturing engineer includes connecting with people as well as machines. Engineers who specialize in manufacturing maintain good working relationships with customers and with workers and supervisors on factory floors. They listen to the suggestions of line operators who work with the equipment and then adjust plans and processes based on these suggestions. Nicole Tacopina, a senior processing engineer who oversees production of medical devices, says, "In a typical work day I will interact with operators, inspectors, quality engineers, managers, supply planning, subject matter experts, and project managers."[10]

On the operations front, manufacturing engineers keep an intense focus on the numerous processes that need to come together to produce a product. They ensure that raw materials are delivered and are safely and efficiently moved to where they are needed on the factory floor. The engineers also spend their workdays observing and adjusting assembly, packing, and shipping processes to make sure all work smoothly.

Most manufacturers have entire departments devoted to each specific process. Engineers spend their days pulling together the needs and desires of each department into a cohesive production strategy while identifying and solving problems along the way. According to manufacturing engineer David Wynn:

> Once the process is fully implemented, and the design is in full production, it becomes the manufacturing engineer's job to look for ways to reduce costs by increasing throughput and improving yield. They'll look at ways to do this while improving safety and [the] quality of their product. A plant is a dangerous place to work, and often times manufacturing engineers work hand in hand with [environmental, health, and safety professionals] . . . to ensure new processes are safe.[11]

Education and Training

Manufacturing engineers are required to have a bachelor of science degree in engineering. They might major in mechanical engineering, industrial engineering, electrical engineering, manufacturing engineering, material science engineering, or industrial engineering technology.

Math is the foundation for all engineering studies, and students should take as many math courses as possible. General math teaches students to think in logical ways, algebra is useful for creating algorithms while writing software, and geometry is used when creating plans and blueprints. Manufacturing technology—which uses electricity, mechanics, heat, light, sound, and optics—is based on applied physics. This means prospective manufacturing engineers should attend physics classes, in which they will learn to grasp how circuits, processors, and other components function.

College coursework for manufacturing engineering majors includes classes in statistics, production systems planning, and manufacturing systems design. Some colleges and universities offer five-year degree programs that provide students with a bachelor's and a master's degree. Graduate coursework might include systems engineering, computer system security foundations,

Manufacturing engineers take part in designing, building, and testing complex production systems. They also plan and set up manufacturing operations and monitor production processes.

probability for electrical engineers, broadband network architectures, and modern active circuit design.

Many companies offer summer intern positions to engineering students who are still in college. Interns are typically paired with mentors and collaborate with others on group projects. Internships offer good networking opportunities and often lead to full-time employment upon graduation. Dominque says she worked as an intern every summer during her college years:

My first internship was with Boeing in Huntington Beach, California. I worked in a materials test lab. It was an amazing experience because of the location and because of the fact that I was jumping right into what I always wanted to do. I was designing, creating, and destroying different materials. I learned processes and gained hands-on experience that a textbook can't give you.[12]

Skills and Personality

Manufacturing engineers work closely with workers on assembly lines, which requires them to combine the talents of a good communicator with those of a tech wizard. The engineers must address—in plain English—questions and problems concerning robots and other machinery, their assembly instructions or operating software, or any issue that might slow the production line.

Attention to detail is an operational skill especially important in the manufacturing field. Engineers work with intricate parts and search for the tiniest glitches in the automation process. The knack for problem solving requires engineers to have an eye for detail when examining physical machinery or lines of computer code. And they also need to be patient and persistent when solving problems. Multitasking skills are important for manufacturing engineers, who need to remain organized and focused while simultaneously dealing with puzzling plans, tiny pieces of hardware, and demands from production managers and chief executive officers. As manufacturing engineer Peter

Raucci makes clear, "Manufacturing engineers require a wide range of skills to be effective in their role. They must be both analytical and people-focused, good communicators as well as technical gurus, and above all, masters of creative problem solving."[13]

Manufacturing engineers who work at big companies or for the government are not required to have a special license. However, those who are seeking a higher level of leadership, with a commensurate level of pay, can benefit from obtaining a professional engineer (PE) license. This requires the engineer to pass a Fundamentals of Engineering exam, work under a licensed PE for four years, and then take a Principles and Practice of Engineering exam.

Working Conditions

In factories that run twenty-four hours a day, manufacturing engineers might work in shifts. Typically, employees on the first shift work from 6:00 a.m. to 3:00 p.m. The second shift runs from 3:00 p.m. until 12:00 a.m., while the third shift overlaps the second, from 10:00 p.m. to 6:00 a.m. The hours worked by manufacturing engineers vary depending on the company. Some work double shifts three or four days a week, while others work one shift for five or six days. Manufacturing engineers might work the same shift all the time, or they may be asked to rotate. They might also be required to work on weekends and holidays.

Employers and Earnings

The Bureau of Labor Statistics (BLS) places manufacturing engineers in the same category as industrial engineers, who earned an average salary of $95,300 in 2021. However, according to the BLS, engineers who work in transportation manufacturing earned several thousand dollars more per year, while those who specialized in computer and electronics manufacturing averaged $99,340 in 2021.

Future Outlook

Manufacturing engineers work in a wide range of industries that produce consumer products. As more companies retool with new robotic equipment, engineers who specialize in manufacturing, research, and development are expected to remain in high demand. According to the BLS, employment of all industrial engineers is expected to grow by 10 percent through 2031.

Find Out More

American Society for Engineering Education (ASEE)

www.asee.org

The ASEE promotes the development of innovative approaches to engineering education. The Education & Careers link on the society's website offers webinars, courses and workshops, and information about academic job opportunities.

Institute of Industrial & Systems Engineers (IISE)

www.iise.org

The IISE is made up of engineering professionals who work in automobile manufacturing, aerospace, heath care, education, and other fields. The institute publishes blogs, white papers, and newsletters and offers training, scholarships, and career advice for young professionals.

Society of Manufacturing Engineers (SME)

www.sme.org

The SME was founded to promote manufacturing technology, attract future engineers, and develop a skilled workforce. The SME Educational Foundation offers training, scholarships, contest and competitions, student summits, and mentorships with professional engineers.

Robotics Engineer

What Does a Robotics Engineer Do?

In 2023 around half of all Americans said they used food-delivery services at least once a month. Around 13 percent reported ordering food for delivery from two to six times a week, according to a report from mobile-based training platform eduMe. The growing consumer reliance on take-out food has created problems for restaurants, which are having trouble finding enough food-delivery workers. The robotics engineers at Kiwibot have found a solution; the company designed and built a fleet of autonomous food-delivery robots equipped with advanced camera sensors, Global Positioning System technology, and artificial intelligence (AI). The little delivery bots can be seen rolling along the sidewalks of twenty-seven college campuses in the United States, avoiding obstacles and planning optimal routes while reducing delivery costs by 65 percent.

Kiwibot is one of several companies revolutionizing the food-delivery business. Postmates similarly employs what it calls Serve delivery robots to transport food, groceries, and small items to customers. Another

A Few Facts

Number of Jobs
About 284,900 in 2021

Pay
Median annual salary of $95,300*

Educational Requirements
Bachelor of science degree in engineering

Personal Qualities
Scientific and technical knowledge, team player, good communication skills, lifelong learner

Work Settings
Offices, drafting rooms, factory floors, and field-testing sites

Future Job Outlook
Growth rate of 2 percent through 2031*

* For all mechanical engineers, including robotics engineers

company, Flytrex, is focused on fast-food flying drone deliveries in North Carolina and Texas.

Autonomous machines that deliver pizza and burritos attract attention, but they make up a small portion of the robots that are used every day. According to robotics engineer Gail Dyer, the greatest demand for engineers is coming from the industrial sector. Dyer says:

One type of Robotics Engineering is . . . where the robots learn their environment and what paths to take. However, I work primarily with industrial robot arms, which is a whole other area of robotics. Some can lift up an entire car body. Some are small enough to sit on a tabletop. . . . The implementation of robots is huge in the industry, since robots are generally not made for a specific purpose. Engineers like me figure out how to implement a robot for my company and for the products we make.[14]

Robotics engineers like Dyer combine several different academic disciplines that focus on mechanical, electronic, and computerized systems. Robotics engineers use mechanical engineering skills to design a robot's moving hardware that includes bearings and actuators (the tiny electrical motor and gearbox that drives a robot's joints). A robot's wiring, control devices, power systems, and sensors require knowledge of electronics engineering. And those who create a robot's processors, programs, and AI systems have experience with computer and software engineering. Then there is the entire field of kinematics. This mathematical discipline, sometimes called the geometry of motion, focuses on the position, velocity, and acceleration of objects and groups of objects. While few robotics engineers are experts in all these fields, they work in teams and combine their knowledge to create sophisticated robotic machinery.

A Typical Workday

Robotics engineers work in offices, in drafting studios, and on factory floors. They spend many hours in front of computers writing code and designing robotic machinery. This generally requires them to use two types of software programs, computer-aided design (CAD) and computer-aided manufacturing (CAM). These programs allow engineers to create complex 3-D drawings and to manufacture parts on 3-D printers. Engineers also write computer code and create software to program robots. Sometimes engineers field-test their robots, and if a robot—such as a delivery robot or police robot—is meant to function outdoors, then engineers watch their creations perform in urban or residential settings, sand, water, and other environments.

Education and Training

Anyone wishing to work designing and building robots needs at least a bachelor of science degree in mechanical, manufacturing, electrical, or industrial engineering. High school students who wish to pursue a career designing, testing, and overseeing production of robotic equipment should focus on STEM subjects. Robotics engineers also make extensive use of physics and chemistry. Computer coding skills are mandatory in robotics. Engineers commonly work with C++, Python, and other programming languages.

Robotics engineers design and build robots that perform a wide array of tasks in many different fields. One such robot is the Kiwibot food-delivery vehicle (pictured).

Beyond technical skills, robotics work relies on teamwork—communications classes like speech and writing help students learn to express their ideas more effectively. In addition to studying math and programming at school, Dyer engaged in extracurricular activities related to the field. "In my freshman year of high school, I went to a Women in Technology camp," she recalls. "We got exposure to a variety of engineering disciplines like Mechanical, Electrical, and Computer Engineering. Then, we got to do robotics! We got to use industrial robots that are used in factories around the world. I fell in love and decided I wanted to be a Robotics Engineer."[15]

College students working toward a bachelor's degree in engineering take courses that focus on pneumatics, hydraulics, CAD/CAM systems, logic, microprocessors, and integrated systems. Some robotics engineers work toward a bachelor's degree in one field, such as mechanical engineering, and a master's degree in another field like electronic engineering. This qualifies them for upper-level positions in the industry.

College students in their sophomore year can apply for co-operative education programs (co-ops) at tech companies or at government research labs like the Jet Propulsion Laboratory in Pasadena, California. Students in co-ops stop taking classes to work full time. The positions are typically paid and last anywhere from three to twelve months. Students receive academic credit and a letter grade for their paid work experience.

Internships also provide numerous opportunities to build experience. Robotics engineer Samuel Sampson states:

> Throughout my seven-year journey to a bachelor's degree, I worked as a software engineering intern a handful of times. Each internship opportunity was through some connection I had made along the way—one time I met software engineers that were regulars at the coffee shop I worked at, one time a professor offered me a job at a company he founded after I did particularly well in his class![16]

Robotics engineers who work at big companies or for the government are not required to have a special license. However, those who work for small businesses might need to become a certified professional engineer (PE). This requires the engineer to pass a Fundamentals of Engineering exam, work under a licensed PE for four years, and then take a Principles and Practice of Engineering exam.

Skills and Personality

Becoming a robotics engineer requires long hours studying the most complex subjects imaginable. Successful engineers have ambition, analytical and scientific skills, and a passion for robotic technology. Engineers need to have excellent communication skills, technical knowledge, and the ability to plan and manage projects effectively and within budget. Rafael Rincón, a robotics engineer at Kiwibot, offers this advice:

> Curiosity has to be your number one quality and you have to be very flexible as well. Robotics involves everything from math and physics to programming and machine learning. You don't have to know everything, but it is very useful to be able to understand all the processes inside

the robot. Most importantly, never forget your social and soft skills. . . . Robotics is all about teamwork, always.[17]

Robotics engineers work in an industry that is constantly changing. This requires robotics engineers to be lifelong learners who stay up-to-date on the latest developments in the field. They remain well informed by reading journals, websites, blogs, and research papers about the latest industry developments, progress in robotics, and efforts of other professionals.

Working Conditions

Robotics engineers might work by themselves in office settings. There they have powerful computers with design software that can simulate robot work environments. Some who work around robotic machinery on production floors are more hands-on. They are required to wear ear protection, goggles, and other safety equipment. Engineers in this field travel to conferences and factories where robotic machines are produced. Those with a master's degree might teach part time or full time.

Employers and Earnings

Traditionally, robotics engineers have found employment at car companies like General Motors and Ford. They also work on autonomous (self-driving) vehicles for large tech companies, including Google and Tesla. Engineers work for major manufacturers of food and consumer goods and design assistive robots that perform a wide range of tasks in the production and packing process. Robotics engineering student Amanda lists other areas where engineering professionals find work: "Robotics Engineers . . . can go into the entertainment industry and work on theme park rides or animatronics, they can help develop and build robots that help in a healthcare setting, either by helping medical professionals or with disability aid, they can work for NASA [the National Aeronautics and Space Administration] and help build rovers that help us explore planets, the opportunities are endless."[18]

The Bureau of Labor Statistics (BLS) does not provide specific job information for robotics engineers. The BLS categorizes them with mechanical engineers, who earned a median salary of $95,300 in 2021. The BLS does say that mechanical engineers who worked in scientific research and development—an area that would include robotics engineers—earned a $102,050 salary on average. While these figures represent average pay, there is great demand for robotics engineers with PhD-level skills in machine vision, deep learning, or machine learning. Those who hold these skills can earn at least $250,000 annually, according to Frank Bertini, robotics business manager at Velodyne Lidar.

Future Outlook

The BLS predicts that the demand for mechanical engineers will grow by only 2 percent through 2031. However, BLS statistics do not consider the fact that demand for robotics engineers is outpacing supply. In 2022 Ann P. Walsh, chief executive officer of Robot.Jobs, reported that on her employment site open positions for robotics engineers increased 500 percent. "[Robotics] careers are in high demand across almost all industries, including industrial, healthcare, biotech, logistics, consumer and more,"[19] she says. This means that the future outlook for robotics engineers is good, and demand should be higher than predicted by the BLS.

Find Out More

Association for Advancing Automation

www.automate.org

This organization provides webinars, certifications, awards, and seminars to engineers, managers, and executives in the robotics industry. Students can find free educational resources such as a beginner's guide to robotics, tech papers, videos, and career advice on the website.

International Federation of Robotics (IFR)

www.ifr.org

The IFR is an international organization that serves the robotics industry through publication of market data, white papers, and blogs and by hosting symposiums and other events. The IFR website provides prospective robotics engineers with learning tools concerning industrial robots, robotics research, and case studies.

Jet Propulsion Laboratory (JPL)

www.jpl.jobs

The JPL is a research and development center for robotic space and earth science missions. The JPL website provides information about hi-tech robotics jobs along with offers of internships and mentorships that would be of interest to prospective robotics engineers.

The Robot Report

www.therobotreport.com

This website covers a wide range of subjects related to robotics engineering, technology, and business. Students interested in robotics can find current events, product information, and research along with webinars and downloadable reports that cover the development, integration, and use of robots.

Mathematician

What Does a Mathematician Do?

Most modern robots would be lifeless hunks of metal and plastic without artificial intelligence (AI) programs. Advanced AI software allows robots to access the Global Positioning System and the internet, process data from sensors, and learn new ways to solve problems. AI provides what some call the "wow factor" in robots, the seemingly magical ability to react to input and take on new tasks. But as mathematician Madiha Jamal writes, "AI is not magic; it's just mathematics. . . . Artificial Intelligence and Mathematics are the two branches of the same tree."[20]

As Jamal makes clear, the most complex robotic machines on earth—and those that explore outer space—would not exist without mathematicians. These professionals work with numerical concepts and use geometry, linear algebra, game theory, computational mathematics, applied mathematics, and other types of math to solve problems in engineering, economics, physics, and other sciences.

Mathematicians who specialize in robotics conduct research concerning movements of robotic arms and legs, and they write software to

A Few Facts

Number of Jobs
About 36,100 in 2021

Pay
Median annual salary of $108,100

Educational Requirements
Bachelor's degree in mathematics, master's degree preferred by most employers

Personal Qualities
Puzzle solver, math lover, analytical, good communicator

Work Settings
Full time in office settings, some travel

Future Job Outlook
Growth rate of 31 percent through 2031

control the movements of the machines. This requires a background in computational number theory, modular arithmetic, calculus, linear algebra, graph theory, and probability. Robotics mathematicians also rely on the knowledge of obscure concepts like Bayesian inference (a technique of statistics) and kinematics, which focuses on the movement—including the position, velocity, and acceleration—of objects and groups of objects. As mathematics professor Syed Zain Mehdi explains, "[Robotics is] all mathematics. It's the foundation that allows us to analyze and define definite rules aiding our aims for robotic design and development . . . [and] to formulate new and innovative concepts."[21]

A Typical Workday

Mathematicians who work in the field of robotics have extremely complex duties. They spend their days analyzing data and developing algorithms for AI programs. They study statistics and computational methods and design encryption systems. This work often involves collaborations with teams of engineers, researchers, physicists, and other highly educated people. Julia Badger, project manager for Robotics and Intelligence for Human Spacecraft, describes her typical day at the National Aeronautics and

Space Administration: "I oversee a team of about 15 engineers and 7 robots, and I manage the technical directions, the schedule and progress, and integrations for our team," she says. "I serve as a subject matter expert for both autonomy and intravehicular robotics for our latest program, Gateway, a . . . space station for human exploration beyond low Earth orbit."[22]

Mathematicians spend many hours conducting and interpreting research and developing projects for their employers. They apply the knowledge they glean to developing algorithms that can be used in real-world projects, such as making robots perform specific functions. A former math professor with the online handle Onzie9 works on AI for a gaming company. He says his employer encourages him to take time off from working on company projects to conduct research on personal interests. The hope is that the mental stimulation of the private projects will lead back to applications relevant to the company's work. He explains:

> Ideally, that research is somewhat related (or possibly relatable) to the work we do, but that can be stretched. In particular, 4 times a year, the entire company takes 3 days to engage in personal projects. After the end of the 3 days, we have the choice to present our work. . . . If I have an idea that is directly related to the work at hand, I can definitely take a couple days to dig up papers, poke around and engage in real research to follow the idea through. I use the designated research days to work on things that are more tangential curiosities.[23]

While side projects can help ease the mentally taxing aspects of their work, mathematicians in the robotics industry commit a lot of time to putting difficult concepts into practical use. Those who spend their days on research need to stay well informed about the latest developments in their field. This means reading mathematical journals and participating in conferences with others in their profession. Mathematicians who work for colleges and universities teach courses while working on research projects. Mathematics professor Eric Platt says, "Ideally [mathematicians] get plenty of money for research grants for the area of research they are in. This means that academic mathematicians are often writing grant proposals and reports on progress. There is also money to be made in writing and editing books, and in consulting."[24]

Education and Training

Some mathematicians have only a bachelor's degree in mathematics, and that is enough to qualify them to work for federal government agencies. However, most employers who hire mathematicians require advanced degrees, either a master of science degree or a doctorate in mathematics.

High school students who hope to become mathematicians need to take as many math courses as possible, including geometry, algebra, statistics, and calculus. Since most math professionals also work with and even create data analysis software, students should become proficient coders familiar with programming languages such as Arduino IDE, Python, and C/C+.

Prospective mathematicians should consider attending summer enrichment programs in math, commonly referred to as math camps. These programs try to make math fun; campers create apps, work on puzzles, participate in competitions, meet professionals in the field, and even build robots. Summer enrichment programs are often hosted by colleges and universities and can be found in most states. The American Mathematical Society website features a detailed list of residential programs nationwide. In addition to boosting their learning, those who participate in math summer camps show that they are motivated students who have an interest in academics. This can enhance college applications.

College students studying for a bachelor's degree in mathematics take advanced math courses such as differential equations and linear and abstract algebra. Most educational institutions require students to take related courses in computer science, en-

Mathematicians who work in the field of robotics analyze data and develop algorithms for artificial intelligence programs. They often work with teams of engineers, physicists, and others.

gineering, or physics. Postgraduate students obtain degrees in theoretical or applied mathematics. Those who wish to go into robotics focus on applied mathematics, which is concerned with solving practical problems.

Master's degree programs in mathematics involve courses in real analysis, complex analysis, probability, scientific computing, and differential equations. Students learn through a combination of lectures and seminars and spend time working independently to solve problems. Those who pursue a doctoral degree in mathematics immerse themselves in extremely complex subjects, including the theory of mathematics, mathematical logic, statistical and mathematical analysis, topology, and stochastic processes.

Skills and Personality

While math can be mind-bogglingly complex, at its most basic level, it is about solving puzzles, and most mathematicians love tackling problems. This requires extremely strong analytical skills. Math professionals spend their days combing through massive amounts of data. They need to remember the smallest details to form analytical opinions. Some mathematicians also need highly developed coding skills because they often write customized software programs to model mathematical formulas and run analyses of specific mathematical problems.

Most mathematicians—obviously—have a deep love for math. Some even see magnificence in math formulas that can equal the aesthetic splendor of a symphony or painting. Mathematics professor Todd Kemp says math can resemble music. He explains, "I find (my) mathematics very musical. There is harmony (and sometimes dissonance) in the way different ideas and structures come together in a mathematical proof. I have even referred to some of the papers I've written as 'arrangements' of compositions by other mathematicians who had greater insight or creativity than I did."[25]

Like most hi-tech workers, mathematicians need to be able to explain their work in plain language to associates, including engineers, scientists, business and marketing managers, and others who are less proficient in math. This is especially true in a field like robotics. A mathematician known as Wintermute93 explains how his soft skills, or people skills, helped him land a robotics job in the defense industry:

> I billed myself not as someone with any one particular competency critical to their business, but as someone who has

a ton of experience taking complex concepts and present-
ing them in an understandable way to a lay audience, with
a . . . desire to learn new things (like, say, whatever the
most interesting technical problems facing their business
are). Soft skills go a long way, you'd be surprised how
many people applying for data-centric jobs are terrible at
communication.[26]

Working Conditions

Most mathematicians have full-time jobs in offices, research centers, and classrooms. But it is commonly said that mathematicians can only concentrate on a task for about three hours before
they need a break from the mental gymnastics of the job. Per-
haps this is why tech companies and other employers often allow
mathematicians to set their own schedules and work from home.

Employers and Earnings

Mathematicians work in many economic sectors, including fi-
nance, insurance, automotive, social media, gaming, and robotics. Some math professionals work in high-end research centers
like the CERN particle physics laboratory in Switzerland or the Jet
Propulsion Laboratory in Pasadena. The Bureau of Labor Statistics (BLS) says the median annual salary for mathematicians was
$108,100 in 2021. Those who worked in the science and tech
sectors, including research and development for AI and robotics,
earned a higher average salary of $129,800.

Future Outlook

Mathematicians are in high demand across a wide range of industries, including robotics. According to the BLS, employment
for mathematicians is expected to grow by 31 percent through
2031. The largest demand for math professionals will come from
the communications and medical manufacturing industries, two

of many growing fields that are relying on statistical analysis to inform business decisions.

Find Out More

American Mathematical Society (AMS)

www.ams.org

The AMS represents mathematicians, researchers, and educators. It promotes mathematics research through the publication of books, journals, and blogs and sponsors programs for student education. The society's website lists information about math summer camps in nearly every state.

Mathematical Association of America (MAA)

www.maa.org

The MAA is an international community of mathematicians, computer scientists, STEM professionals, students, and teachers. The association hosts competitions, provides curriculum resources, and offers student resources, including problem-solving books and information about regional conferences.

Robotics Education & Competition (REC) Foundation

https://roboticseducation.org

The REC Foundation was founded to increase student interest in science, technology, engineering, and math. The foundation provides programs, workshops, and contests focused on robotics and other innovative technologies and sponsors the Girl Powered program to mentor female students.

Society for Industrial and Applied Mathematics (SIAM)

www.siam.org

This professional society caters to mathematicians and researchers in business, government, and science. SIAM provides students access to videos, handbooks, competitions, and information about summer schools, internships, and careers in math.

Computer and Information Research Scientist

What Does a Computer and Information Research Scientist Do?

While new apps are launched every day, few make headlines. But when the ChatGPT app dropped in late 2022, the world took notice. The chatbot—which uses artificial intelligence (AI) to create songs, stories, scripts, photos, student essays, and more—was downloaded over 100 million times by January 2023. This made ChatGPT the fastest-growing software application in history.

After ChatGPT launched, there was a great debate among journalists and average users about the pros and cons of using AI to take tests and write essays. But there was little discussion about the people who created ChatGPT. Like all AI software, the program was initially developed by computer and information research scientists. Depending on their job, these professionals might also be referred to as computer scientists, research scientists, applied scientists, or AI scientists. Those who specialize in robotics

A Few Facts

Number of Jobs
About 33,500 in 2021

Pay
Median annual salary of $131,490

Educational Requirements
Master's degree

Personal Qualities
Math and computer skills, analytical, good communicator, detail oriented, team player

Work Settings
Full time in offices, laboratories, and workshops

Future Job Outlook
Growth rate of 21 percent through 2031

focus on AI and machine learning, which allows robots to teach themselves to make decisions without further human input. Chatbots are robots, though like many AI programs, they lack a physical form. And while some AI scientists do work on software implanted in physical robots, others concern themselves with chatbots and interactive programs based on the internet or within downloadable applications.

Because of the varied industries using AI, computer and information research scientists help engineers design games and gaming hardware, navigation and guidance systems, smartphones, self-driving cars, and other products. AI scientists who work in the entertainment industry are responsible for the AI programs that suggest music and movies to users of Netflix, Spotify, and other streaming services. Anyone who shops e-commerce sites interacts with the work of AI scientists. These professionals perfect the website's ability to mine data, translate languages, analyze hardware and software performance, address customer complaints, prevent fraud, and improve core search functions.

Some of the most important work performed by computer and information research scientists involves writing, testing, and analyzing complex machine learning algorithms. AI consultant Raman Arora explains, "Machine Learning algorithms [are] the point where real AI starts. Using these algorithms you can improvise your output each time by making your model . . . remember its mistakes and not repeat them."[27]

A Typical Workday

Computer and information research scientists spend much of their time reading and learning. They might seek new ways to do something by studying written hardware and software manuals. Dissertations and academic research papers provide information about complex subjects such as logic, information theory, and topology (the properties of geometric objects). Researchers also consult older books and essays that cover traditional subjects, including philosophy, math, and physics. Some spend their time as AI promoters. In this role they write blogs and articles aimed at teaching the public about machine learning and the benefits of AI.

AI researcher Rik Koncel-Kedziorski says his typical workday consists of three main activities: learning, innovating, and reflecting. He explains:

> The "learning" phase of research would be reading research papers, whereas the "innovating" phase mostly involves talking to people, but sometimes requires writing math equations. I would say I spend roughly 85% of my time learning and the other 15% innovating, but that ignores the significant portion of my day spent reflecting: digesting or recuperating from the work of learning and innovating. I cannot stress enough the importance of unstructured reflection time in the psychic life of a researcher![28]

Koncel-Kedziorski says he tries to read papers on challenging concepts or innovative technology every day. These papers test his preconceived notions or provide new information that can be used to solve problems. He also learns by talking to researchers in other fields who might provide a fresh look about a difficult subject. As Koncel-Kedziorski makes clear, working on high-level technical research requires scientists to reflect on various subjects, both theoretical and practical. On one level they think about big-picture ideas—how they can improve a customer's or business user's experience. On the other hand, AI research scientists meditate on real-world applications. They think of ways they can tweak software or solve a specific technical problem. Answers often present themselves at odd moments when the researcher is not at work. Koncel-Kedziorski attributes these solutions to what he calls brain magic. "I will be washing dishes or sailing the boat on a weekend," he explains, "and meanwhile my brain is busy synthesizing what was learned and innovated in the last week all on its own. Regularly, I will be struck with a powerful new insight or direction for future research while in the middle of some non-work activity and must quickly jot it down in my Notes app for future consideration."[29]

In addition to deep thinking, computer and information research scientists function as team leaders in the workplace. They schedule and attend meetings, hire and fire employees, evaluate project performance data, and create proposals for new projects. Researchers are responsible for explaining project goals, policies, and projects to coworkers and management. They might also participate in multidisciplinary projects at universities or research labs.

Education and Training

Computer and information research scientists typically have degrees in computer engineering, software engineering, information systems, information technology, or a related field. While a bachelor's degree might be adequate for entry-level jobs in the field, most employers require information research scientists to have a master's degree. Those who wish to conduct pure research need a doctoral degree.

Becoming a computer and information research scientist is a long-term goal that requires education, experience, and determination. High school students who hope to work as computer research scientists one day should focus on math courses, including calculus, statistics, trigonometry, and algebra. Students should take physics, chemistry, and communications as well as computer science classes. Coding is central to a career as a computer and information research scientist, and students can learn important programming languages like R, Python, and SQL online.

Those pursuing a bachelor's degree take courses in programming, software development, math, and coding. Students study linear algebra, optimization, neuroscience, cognitive science, algorithms, and the theory of computation. Internships provide an opportunity to observe professional computer and information research scientists at work. Interns gain valuable experience managing and troubleshooting problems with computers, software, robotic equipment, and other technology. While working as an intern, students can expand their networks, gaining contacts with those who can provide references and job recommendations after graduation.

Skills and Personality

Computer and information research scientists develop technical knowledge of circuit boards, processors, chips, computer hardware and software, and other electronics. Scientists in this field rely on analytical skills to troubleshoot software and equipment and find innovative ways to solve problems. They use logic and reasoning to identify the strengths and weaknesses of the solutions they devise to fix problems.

Computer and information research scientists are often team leaders who need highly developed soft skills. The scientists work with analysts, engineers, project managers, department heads, and vendors. They need to listen to coworkers and clearly present information and ideas to others.

Computer and information research scientists should also understand the business aspects of their work. Their employers likely spend large sums of money on research, and the scientists need to make their work pay off at some point in the future.

Coffee, Coding, and Reading

"All days start with . . . coffee, individual contribution work, catered lunch at desk, continuation of work until the task is finished, with another coffee in between. What I'm working on differs depending on the stage of the product cycle. If I'm on product development tasks, I'm doing a lot of software engineering work like coding, writing documents, code reviews, preparing demos, etc. If it's a research task, I'm reading papers, studying code bases, prototyping, running experimentations, tuning the models, evaluating performance, [organizing] the project for sharing and reproducibility, discussing the results, and preparing presentations and writing papers."

—Matineh Shaker, applied scientist

Quoted in DeepLearning.AI, "Working AI: At the Office with Applied Scientist Matineh Shaker," 2023. www.deeplearning.ai.

Working Conditions

Most computer and information research scientists work forty hours per week in offices, laboratories, and workshops. When major projects are nearing completion and deadlines are looming, they might work overtime, including nights and weekends. With the computer, software, and robotics industries changing very rapidly, computer and information research scientists regularly toil to update their knowledge and skills. They take online courses and might even attend college classes. Traveling to attend seminars is another aspect of the job. Conferences allow researchers to meet and socialize with other experts in their field. These connections can be very important when advice is needed for particularly difficult projects.

Employers and Earnings

The Bureau of Labor Statistics (BLS) says that 31 percent of computer and information research scientists work for the federal government. They might work in cybersecurity or hold a top secret clearance at the US Department of Defense, where they develop weapons systems. Other computer and information research scientists are employed in the tech and biotech industries and at engineering firms.

According to the BLS, median annual salary for computer and information research scientists in 2021 was $131,490. Those who were involved in designing computer systems, which would include robotics, earned the highest pay, around $161,870.

Future Outlook

According to the BLS, computer and information research scientists are in high demand; employment for these professionals is expected to grow by 21 percent through 2031. Work in AI is helping drive this growth.

Find Out More

Association for the Advancement of
Artificial Intelligence (AAAI)

https://aaai.org

The AAAI is a scientific society focused on emulating thought and intelligent behavior in computers. Artificial intelligence is spreading into information technology (IT) applications, and the AAAI offers conferences, workshops, periodicals, and books for prospective IT managers. The organization also provides student scholarships, grants, and other honors.

Computing Research Association (CRA)

https://cra.org

The CRA is dedicated to linking computer researchers across industry, academia, and government. The association has a strong focus on students, and its website provides information about research grants, awards programs, grad school options, and career building.

IEEE Computer Society

www.computer.org

The IEEE Computer Society is dedicated to computer science and technology and provides information about networking and career-development for educators, scientists, engineers, and students. Its website contains numerous educational and career-building resources, including online courses and books, webinars, scholarships, and certification preparatory information.

USENIX Association

www.usenix.org

USENIX is also known as the Advanced Computing Systems Association. It is a community of engineers, systems administrators, scientists, and technicians that hosts advanced computing conferences, promotes research, and shares information. The USENIX student section provides tech sessions and tutorials, grants, and information about student awards.

Data Scientist

What Does a Data Scientist Do?

Data is a simple word that comprises facts, information, and statistics related to anything and everything. Anyone who uses a smartphone, computer, tablet, or other digital device generates lots of data every day. People create data every time they stream a video, order food or a ride, check the weather, pull up a map, text their friends, or do anything else online. With nearly 5 billion active internet users worldwide, 329 million terabytes of data are created each day, according to the online data gathering platform Statista.

Around 90 percent of the world's data was generated in the past two years alone. This mind-boggling amount of information is a digital gold mine that drives the internet economy, which was worth more than $2.4 trillion in the United States in 2021. Businesses and organizations that want to earn profits from the digital economy rely on data scientists to capture and analyze this information. Data scientists search for patterns in databases and analyze them to help companies make business decisions. Data scientist Rashi Desai describes the job this way: "It is about the extraction of knowledge

A Few Facts

Number of Jobs
About 113,000 in 2021

Pay
Median annual salary of $104,123

Educational Requirements
Master's degree in data science, computer science, engineering, mathematics

Personal Qualities
Highly developed science and math skills, business acumen, good communicator, team player

Work Settings
Full time in offices

Future Job Outlook
Growth rate of 36 percent through 2031

from data to answer a particular question. For me, putting it simply, data science is a power that allows businesses and stakeholders to make informed decisions and solve problems with data."[30]

To understand the job of a data scientist, it is helpful to look at an e-commerce site like Amazon, which is responsible for about 40 percent of all online sales in the United States. Amazon maintains vast databases filled with information on its customers that includes names, ages, phone numbers, addresses, credit card numbers, shopping histories, and even religious and political affiliations. Data scientists who work for Amazon study information about customers' buying histories, product returns, seasonal purchases, and other factors. The retailer can use this data to time the release of new products and offer deals and sales. The data also predicts economic upturns and downturns and can help the company determine when to hire and fire employees.

The job search website Glassdoor names data scientist as one of the top jobs in America every year. Data scientists earn big salaries and are said to have high levels of job satisfaction. These facts drive a growing number of college students to enroll as data science majors. But data science is not for everyone. Those who work in the profession need to possess outstanding scientific, mathematical, and technical skills that can only be developed through hard work and experience.

Data scientists in the field of robotics gather, examine, and shape data generated by artificial intelligence (AI). This requires them to focus on statistics, mathematical algorithms, and massive sets of structured and unstructured data, seeking patterns or trends that affect a robot's performance. They develop and test new algorithms and develop predictive models meant to prevent future problems.

Driverless cars, which are basically robots with passenger seats, provide a good example of a data scientist's work. A self-driving machine generates around 4,000 gigabytes of data each day. The data is provided by its Global Positioning System, video cameras, cloud-based maps, and numerous sensors that detect objects, lane markings, and the positions of other vehicles. Data scientists analyze and interpret all that data to "teach" the car to interact with the physical world. Data scientists invent highly detailed software programs that help the car navigate roads without incident.

Data scientists working with autonomous vehicles focus on five big issues: perception, localization, prediction, planning, and control. Perception helps the car "see," localization lets it know exactly where it is, prediction helps it identify potential hazards, and planning and control keep the car moving safely forward. Programming these processes into a robotic brain requires data scientists to use concepts of machine learning. In this process the robot constantly collects data, interprets it, and updates its algorithms. As data scientist Deepak Dutt writes:

> With data being the fuel for self-driving cars, it would not be a mistake to say that Data Science is the driving force behind autonomous vehicles. Though self-driving cars are still far from making accurate decisions like a human mind, data scientists are working hard to improve the technology. Thanks to Data Science, Machine Learning, and Deep Learning algorithms, the day is coming closer when self-driving vehicles will be compatible with humans and save more human lives by reducing road accidents.[31]

A Typical Workday

Mining data to create new and improved technology is only one part of a data scientist's job. According to the business analytics company Anaconda, data scientists spend around 40 percent of their time preparing and cleaning data. Data cleaning is necessary because data is often corrupt, inaccurate, incomplete, irrelevant, or in the wrong format. Most data scientists say this is the least enjoyable part of their job. Once the data is cleaned and standardized for practical use, data scientists mine the data in search of patterns to analyze. During this process they refine algorithms and build training sets for machine learning. These tasks require long periods of concentration, working alone with minimal interruptions. On the positive side, data scientists can work wherever they have the computing power to perform their jobs. Some avoid commutes and other hassles by working from home.

Education and Training

Most data scientists have a master's degree in data science, computer science, engineering, mathematics, or a related field. But according to Desai, "Data Science is essentially about programming."[32] Students who hope to become data scientists should be able to understand programming languages like Python, R, MATLAB, and SQL. Many colleges and universities offer data science boot camps—four- to six-month crash courses in coding, machine learning, data science, and other relevant topics. Students should also take as many math courses as they can and study database theory, database design, and operating systems like Windows and Linux.

After high school, prospective data scientists will earn a bachelor's degree in data science, statistics, physics, applied business analytics, or a related subject. Most students then go on to earn a master's degree. And around half of data scientists have a PhD, which gives them the educational background necessary to work in advanced robotics.

Skills and Personality

Data scientists are highly organized people with skills in mathematics and analytics. They have strong powers of concentration, an eye for detail, and the ability to work alone with mind-bending piles of data day after day. Data scientists are also extremely competent coders who work with one or more programming languages.

In addition to technical and scientific talents, data scientists need good business sense. They must understand the industry they are working in and what types of problems their employer needs to solve.

Data scientists also need to work efficiently in a business environment. They participate in company meetings and are required to explain their complex data analyses to engineers, marketers, operations managers, chief executive officers, and others. This means they must communicate clearly and in plain language so all can understand. Data scientist Adam Kovarik writes:

[Coworkers] have decisions to make and a bottom line to impact. You are in that room giving that presentation for one simple reason: to improve the business. . . . Those people [in the meeting] want the tools and information at their fingertips to make the business better. Focusing on the minutiae may sell your intellect, but it may also disconnect you from the true goals of the organization. Once this happens, your analysis becomes a footnote.[33]

Kovarik says that listening is as important as talking; data scientists need to listen closely to others in order to understand company objectives and to ensure that others recognize the manner in which the data at hand can help reach those goals.

Working Conditions

Data scientists work in offices with powerful computers that access servers where trillions of bytes of data may be stored. Data scientists often are trying to solve complex problems by poring over the data and drawing useful conclusions. They also attend meetings to work on strategies with company executives, develop new products with engineers, and create ad campaigns with marketers.

Employers and Earnings

Nearly every business makes important decisions based on information gleaned from digital sources. Data scientists are the professionals who make sense of this information, and they work in almost every business sector. Data scientists are employed by major online retailers, work for social media companies including TikTok and Instagram, and analyze data for banks, insurance companies, and the health care industry. Those who specialize in robotics are employed by manufacturers of almost every mass-produced product, and they work for auto and tech companies striving to perfect self-driving cars. Wherever data scientists work, they can expect to earn six figures. According to Glassdoor, the median salary for data scientists was $104,123 in 2023. The website also said that data scientists could expect more than $23,000 in additional pay from bonuses, commissions, and profit sharing.

Future Outlook

Because the economy, education, and government depend on data in the digital age, there is a strong need for data scientists. The Bureau of Labor Statistics says the demand for data scientists is expected to grow at the rapid pace of 36 percent through 2031, much faster than the average for all occupations.

Find Out More

Association for Computing Machinery (ACM)

www.acm.org

The ACM is a professional organization for data scientists, computing educators, and researchers. The association has over eight hundred local chapters worldwide open to students and professionals. The ACM website offers educational resources for students, including tech talks, journals, and research papers.

National Center for Women & Information Technology (NCWIT)

https://ncwit.org

This organization works to increase the participation of girls and women in computer science fields. The NCWIT is focused on students and offers a wide range of educational resources, including videos, blogs, and a newsletter.

TDWI Upside

https://tdwi.org/pages/upside.aspx

This website is aimed at data scientists and those who are considering entering the field. TDWI provides workshops and on-site courses on data modeling, business analytics, advanced data science, and leadership and management. Research resources include webinars, e-books, reports, and infographics.

Source Notes

Introduction: Rapid Growth in Robotics

1. Quoted in John Roach, "The Quest to Deploy Autonomous Robots Within Amazon Fulfillment Centers," Amazon Science, October 24, 2022. www.amazon.science.

Robotics Programmer

2. Sam Daley, "Robotics Technology," Built In, August 18, 2022. https://builtin.com.
3. Quoted in Quora, "How Does Programming for Robotics Differ from Programming in General?," May 13, 2023. www.quora.com.
4. Quoted in Sebastian Trella, "Short Interview with Roy, the Founder of 'the OffBits,'" *Robots Blog*, April 5, 2023. https://robots-blog.com.
5. Quoted Chris Stewart, "Interview: A Day in the Life of a Programmer," *DevMio Blog*, September 6, 2019. https://devm.io.
6. Devin Partida, "Hard & Soft Skills That Can Improve Your Robotics Education," *Robotlab Blog*, 2023. www.robotlab.com.

Manufacturing Engineer

7. Dominque, "A Day in the Life of Manufacturing Engineer Dominique," *All Together* (blog), Society of Women Engineers, February 23, 2021. https://alltogether.swe.org.
8. Quoted in Mars International, "A Day in the Life of a Manufacturing Engineer," November 23, 2016. www.marsint.com.
9. Joe Dyer, "A Day in the Life of a Manufacturing Engineer," Disher, August 17, 2017. https://disher.com.
10. Nicole Tacopina, "A Day in the Life of a Manufacturing Engineer: Nicole Tacopina," *All Together* (blog), Society of Women Engineers, January 6, 2023. https://alltogether.swe.org.
11. Quoted in Quora, "What's It like to Work as a Manufacturing Engineer?," 2019. www.quora.com.
12. Dominque, "A Day in the Life of Manufacturing Engineer Dominique."
13. Peter Raucci, "Meet SPM Manufacturing Engineer—Peter Raucci," SPM, 2022. www.spmswiss.com.

Robotics Engineer

14. Gail Dyer, "A Day in the Life of a Robotics Engineer," *All Together* (blog), Society of Women Engineers, June 11, 2020. https://all together.swe.org.

15. Dyer, "A Day in the Life of a Robotics Engineer."

16. Samuel Sampson, "Student Spotlight: Samuel Sampson," Georgia Tech, 2023. https://omscs.gatech.edu.

17. Quoted in María José Guzmán Rodriguez, "A Day in the Life of AI & Robotics Engineers at Kiwibot," Kiwibot, May 31, 2022. www .kiwibot.com.

18. Quoted in Allison Osmanson, "Robotics Engineering Student Spotlight: Amanda, Courtney and Gemma," *All Together* (blog), Society of Women Engineers, April 29, 2022. https://alltogether.swe.org.

19. Quoted in Automation.com, "2022 Trends from Robots.Jobs Show Dramatic Growth in Robotics and Artificial Intelligence Career Opportunities," February 7, 2022. www.automation.com.

Mathematician

20. Madiha Jamal, "Why Is Mathematics Vital to Thrive in Your AI Career?," *Towards Data Science* (blog), January 15. 2020. https://to wardsdatascience.com.

21. Quoted in Quora, "What Are the Math Prerequisites for Robotics?," January 24, 2018. www.quora.com.

22. Quoted in Hollie Jaques, "Interview with Julia Badger, Project Manager for Robotics and Intelligence for Human Spacecraft at NASA," Re-Work, 2023. https://blog.re-work.co.

23. Quoted in Reddit, "What Do Mathematicians Actually Do on a Day-to-Day Basis?," 2022. www.reddit.com.

24. Quoted in Quora, "What's Life like as a Pure Research Mathematician?," 2022. www.quora.com.

25. Quoted in Michelle Franklin, "Rhythm and Proofs," UC San Diego Today, May 1, 2023. https://today.ucsd.edu.

26. Quoted in Reddit, "What Kind of Industry Jobs Are Out There for Pure Mathematicians?," 2023. www.reddit.com.

Computer and Information Research Scientist

27. Quoted in Quora, "Where Do I Start a Career Path in AI?," May 6, 2017. www.quora.com.

28. Rik Koncel-Kedziorski, "A Day in the Life of an AI Researcher," *Kensho Blog*, May 16, 2023. https://blog.kensho.com.

29. Koncel-Kedziorski, "A Day in the Life of an AI Researcher."

Data Scientist

30. Rashi Desai, "Top 10 Skills for a Data Scientist," *Towards Data Science* (blog), January 2, 2020. https://towardsdatascience.com.
31. Deepak Dutt, "How Is Data Scientist Driving Autonomous Cars?," SkilloVilla, August 10, 2021. https://blogs.skillovilla.com.
32. Desai, "Top 10 Skills for a Data Scientist."
33. Adam Kovarik, "Beyond the Technical: Connecting Data Skills to Business Skills," StatTrak, July 1, 2019. https://stattrak.amstat.org.

Interview with a Robotics Engineer

Julia Badger has been working in space robotics for fourteen years. She is currently the autonomous systems technical discipline lead at the National Aeronautics and Space Administration (NASA). In this role Badger serves as the project manager for the Lunar Gateway, a space station that will orbit the moon. She answered the following questions by email.

Q: Why did you decide to become a robotics engineer?

A: I knew that I wanted to work in robotics, so I looked at the two degrees that were most prevalent in that field—mechanical and electrical engineering. I decided on mechanical engineering because it seemed more versatile—I had to take classes in electrical engineering, materials, and aeronautics, to name a few.

Q: How does your degree in mechanical engineering relate to your work in robotics?

A: My robotics work focused on developing algorithms to allow robots to do tasks somewhat autonomously, or independently from human supervision. It involves writing software to interact with the environment (using sensor values to generate actuator commands) to make the robot function as intended. The basic concept of system control governs this work and was a key part of my mechanical engineering courses.

Q: Can you describe your typical workday (or night)?

A: Most engineering projects are team sports, and robotics is no different. One of the most important parts of any engineer's day is communicating with his or her teammates. So, email, side conversations, and meetings are a typical part of most workdays. There is also typically time alone with your computer, working on your project (writing code, designing an algorithm, or doing analysis).

Some days also include some testing, particularly if you get to test your work on the actual robot. On the best days, your testing works!

Q: What do you like most about your job?
A: I now get to lead teams of engineers who are making amazing things happen. My favorite part of my job is seeing my ideas become reality. I love it when I'm able to help a group of people wrap their heads around a problem and see them build a solution to it.

Q: What do you like least about your job?
A: My least favorite part of my job is dealing with budget. Engineers get to solve really interesting problems, but it's not enough to just come up with an idea—the idea needs to work within a certain amount of time and within a set amount of money. These realities make the problem more fun to solve, but having to think about them isn't my favorite!

Q: What personal qualities do you find most valuable for this type of work?
A: I look for engineers who can work well as part of a team. This means good communication skills, the ability to both generate ideas and support other people's ideas, and a willingness to do tasks that might be a bit outside of their comfort zones. Engineering work can be very difficult, and it can take a long time to see the results of your work. The ability to work hard for delayed gratification is extremely important for any STEM career. Creativity is also very important for engineering development tasks! Engineers are problem solvers, so being able to think about a problem in a new way is a very important trait to have.

Q: What advice do you have for students who might be interested in a career as a mechanical engineer?
A: I suggest taking all the science and math that you can. Involve yourself in groups, activities, and classes that have team projects, regardless of the nature of the project. Spend time using your

hands to build things. Watch how people around you solve problems. If you are good at identifying problems, you might be a budding entrepreneur!

Q: What are your current projects?

A: I recently spent many years working on Robonaut, a humanoid space robot that spent years on the International Space Station learning how to maintain spacecraft systems with limited human supervision. My latest project is developing a brand-new system for NASA's Lunar Gateway space station. The Gateway is part of the Artemis campaign that will establish a sustained human presence on the moon. The Gateway will allow humans to enter lunar orbit, gather supplies from logistics spacecraft, and transfer vehicles meant to either go to the Moon's surface or the Earth's surface. The Gateway will spend months or years in space between human visits and is meant to operate autonomously for up to 3 weeks, independent of Earth-bound Mission Control. As a result, a Vehicle Systems Manager (VSM) is needed to provide integrated mission control functions for the Gateway. This autonomous software control system will provide mission and timeline management, resource management, fault management, and integrated vehicle control and operations for Gateway, keeping the station safe and operational between virtual visits from the Earth-based Mission Control Center operations teams. . . . This work is a culmination of many projects that I and other engineers have been researching and developing for decades and will be the first time ever that many of these technologies will fly on human spacecraft. The VSM represents a new operational paradigm for human spaceflight, and I'm honored to be leading the team developing it for our upcoming launch of Gateway.

Other Jobs in Robotics

Aerospace engineering and operations technician

Assembler and fabricator

Computer and information systems manager

Computer hardware engineer

Computer programmer

Computer systems analyst

Corporate technology officer

Cost estimator

Database administrator

Electrical engineer

Electronics engineer

Industrial engineer

Industrial engineering technician

Industrial production manager

Information security analyst

Logistician

Machinist

Management analyst

Materials engineer

Mechanical engineer

Mechanical engineering technician

Network and computer systems administrator

Operations research analyst

Physicist

Quality control inspector

Software developer

Solderer

Statistician

Systems engineer

Welder

Editor's note: The online *Occupational Outlook Handbook* of the US Department of Labor's Bureau of Labor Statistics is an excellent source of information on jobs in hundreds of career fields, including many of those listed here. The *Occupational Outlook Handbook* may be accessed online at www.bls.gov/ooh.

Index

Picture Credits

About the Author

Stuart A. Kallen is the author of more than 350 nonfiction books for children and young adults. He has written on topics ranging from the theory of relativity to the art of electronic dance music. In 2018 Kallen won a Green Earth Book Award from the Nature Generation environmental organization for his book *Trashing the Planet: Examining the Global Garbage Glut*. In his spare time he is a singer, songwriter, and guitarist in San Diego.